TIME for KiDS

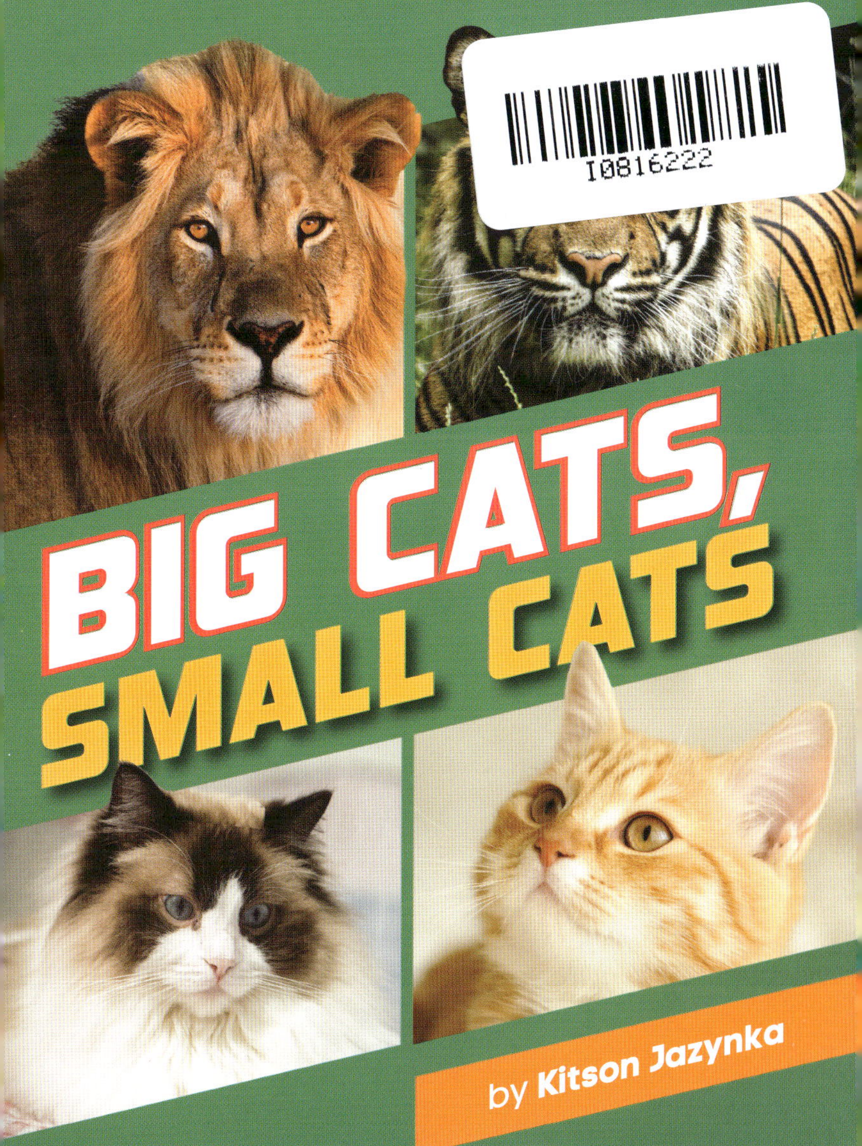

BIG CATS, SMALL CATS

by Kitson Jazynka

PENGUIN YOUNG READERS LICENSES
An imprint of Penguin Random House LLC
1745 Broadway, New York, New York 10019

First published in the United States of America by Penguin Young Readers Licenses,
an imprint of Penguin Random House LLC, 2025

Visit us online at penguinrandomhouse.com.

Library of Congress Cataloging-in-Publication Data is available.

Manufactured in China

ISBN 9780593891407 10 9 8 7 6 5 4 3 2 1 TOPL

Design by Leilani Abdur-Razzaq and Abby Dening

Photo credits: Cover: Getty Images: (Aegean) Caíque de Abreu/iStock, (background) 1xpert/iStock, (cheetah) LuCaAr/iStock, (lion) Seyms/iStock, (orange cat) elenaleonova/E+, (ragdoll) peredniankina/iStock, (Siamese) Tsuji/iStock, (tiger) Wirestock/iStock, (yarn) Alexandrum79/iStock. **Interior: Alamy:** 7: (top) Pardofelis Photography; 15: Tierfotoagentur/J. Hutfluss; 24: (top) Marc Henrie/DK/Dorling Kindersley ltd; 27: (top) blickwinkel, (bottom) Diego Grandi; 76: (top) WireStock/Wirestock, Inc.; 78: (bottom) Arco/G. Lacz/imageBROKER.com GmbH & Co. KG. **Getty Images:** 1: (lion) Seyms/iStock, (orange cat) elenaleonova/E+, (ragdoll) peredniankina/iStock, (tiger) Wirestock/iStock; 2–3: BirdImages/E+; 5: Leoba/iStock; 6: (top) AlexKane/iStock, (bottom) Wirestock/iStock; 7: (bottom) Michael Kloth/Corbis; 8: (top) 1827photography/iStock, (bottom) Wirestock/iStock; 9: (top) slowmotiongli/iStock; 10–11: Kilito Chan/Moment; 12: (top) Volchanskiy/iStock, (bottom) Jolkesky/iStock; 13: (top) Wirestock/iStock, (bottom) Tambako the Jaguar/Moment Open; 14: Wirestock/iStock; 16–17: Seregraff/iStock; 18: (top) Vadimborkin/iStock, (bottom) slowmotiongli/iStock; 19: (top) Kevin Schafer/The Image Bank, (bottom) Thomas Leirikh/iStock; 20: (bottom) GlobalP/iStock; 21: (top) peterotoole/iStock, (bottom) Bolot/E+; 22–23: Zuzule/iStock; 24: (bottom) mlorenzphotography/Moment; 25: (top) StuPorts/iStock, (bottom) Ken Griffiths/iStock; 26: (top) cunfek/iStock, (bottom) Milk Photography/iStock; 28–29: Mateo Juric/iStock; 30–31: Rufous52/iStock; 32: (bottom) Okssi68/iStock; 33: (top) Rachelle Vance Photography/Moment Open, (bottom) Willowpix/iStock; 34: Todorean Gabriel/iStock; 35: Yolya/iStock; 36–37: Phichaklim1/iStock; 38: Donyanedomam/iStock; 39: Wayne Marinovich/iStock; 40: (top) GlobalP/iStock, (bottom) JMrocek/iStock; 41: Arx0nt/iStock; 42–43: Sonja Probst/500px; 44: Warmlight/iStock; 45: (top) harryjco/iStock; 46: (top) SusanLeggett/iStock, (bottom) Kryssia Campos/Moment; 47: StockPhotoAstur/iStock; 48–49: Palenque/iStock; 50: (top) slowmotiongli/iStock, (bottom) NancyAyumi/iStock; 51: Colin Langford/500px/500Px Plus; 52: (top) undefined undefined/iStock, (bottom) Nynke van Holten/iStock; 53: (top) Nils Jacobi/iStock, (bottom) Erich Schmidt/imageBROKER; 54–55: Mint Images/Art Wolfe/Mint Images RF; 56–57: MaggyMeyer/iStock; 58–59: TRAVELARIUM/iStock; 60: slowmotiongli/iStock; 61: Jupiterimages/PHOTOS.com; 62: (top) Kosin_Sukhum/iStock, (bottom) Juan Carlos Vindas/Moment; 63: tashka2000/iStock; 64–65: Astrid860/iStock; 66–67: Martin Ruegner/Stone; 68: mihail1981/iStock; 70–71: Kathleen Reeder Wildlife Photography/Moment; 72–73: Chase Dekker Wild-Life Images/Moment; 74–75: Natalia Lebedinskaia/Moment; 76: (bottom) Senchy/Moment; 77: slowmotiongli/iStock; 78: (top) boryak/iStock; 79: (top) VictoriaBee/iStock, (bottom) Lucy Lambriex/Photodisc; 80–81: Kyslynskyy/iStock; 82: (top) Kacy Kizer/Moment, (bottom) Bob de Graffenried/500px; 83: (top) Nynke van Holten/iStock, (bottom) Iuliia Alekseeva/iStock; 84–85: abzerit/iStock; 86: (top) Wirestock/iStock, (bottom) Picture by Tambako the Jaguar/Moment; 87: davit85/iStock; 88: (top) Whitworth Images/Moment, (bottom) Chris Daniels/iStock; 89: AbElena/iStock; 90–91: Jonas Stenqvist/500px; 92: (top) Evrymmnt/iStock, (bottom) slowmotiongli/iStock; 93: Koraysa/iStock; 94: Jenar/iStock; 95: Volchanskiy/iStock. **Hsiao-Pin Lin:** 69. **Wikimedia Commons:** 9: (bottom) via Aminet, collection © Urban D. Mueller; 20: (top) Jim Sanderson (CC BY-SA 3.0); 32: (top) Larry Johnson/@johnsonphotography (CC0); 45: (bottom) Mauro Tammone (CC BY 3.0).

CONTENTS

INTRODUCTION

Cats—big and small, wild and domesticated—are among the most beloved animals in the world. Over the millennia, cats have purred, chirped, chatted, snuggled, and headbutted their way into the hearts of humans everywhere. Around ten thousand years ago, people settling on the shores of the Mediterranean Sea stored large amounts of grain. This attracted rodents. Small wildcats came to hunt, thus solving the pest problem. This started the process of domestication, in which humans and cats happily lived together and made each other's lives a little easier.

Since ancient times, humans have documented their love for cats. Ancient Egyptian artworks depict cats. Cats appear in folklore throughout history. Big cats are revered as symbols of power, royalty, and strength. In modern times, cats have starring roles in books, movies, TV shows, and social media accounts.

American households today are home to about sixty-one million pet cats. They comfort us, make us laugh, and delight us with their antics and curiosity. Scientific research has even shown that petting a cat releases a flow of "love" hormones in our bodies, making us feel better. In this book, we'll dig our claws into finding out about the many kinds of cats. We hope you'll learn a bit about your favorites—and discover some new ones, too.

We chose a wide variety of cats—from tigers to lounging long-haired house cats and even one very unusual domestic cat bred to look like a wolf. Some of them may seem very different from one another. But scientists consider cats to have a "conservative design." This means that even though there are differences in size, color, and lifestyle, all cats are pretty much built the same, with round heads, sharp teeth, and tails.

As you read, think about how lucky we are to share a planet with these incredible animals—some wild, some not. Many of the species you'll learn about are endangered, or close to becoming extinct in the wild. Scientists, conservationists, and even regular people are working hard to save them. Kids can join in on efforts to protect endangered big cats by learning as much as they can and sharing what they know. Kids can help domestic cats, too. It's a great idea to consider adopting a rescue cat before buying from a breeder, for example. Millions of adorable cats are out there right now waiting for loving forever homes.

ABYSSINIAN

Some experts believe that the regal **ABYSSINIAN** is the domestic breed considered sacred by ancient Egyptians. Others believe the breed originated in Southeast Asia. Its eyes can be hazel, green, or gold. These affectionate and quiet kitties closely resemble the African wildcat. No matter where they're from, this breed is supersmart and loves to learn tricks such as "go fetch."

AEGEAN

The **AEGEAN** cat is a rare breed hardly ever seen outside of its native islands in Greece. They make chatty, friendly pets and have a history that dates back thousands of years. Archaeologists have discovered evidence that these cats lived on the island of Cyprus about ten thousand years ago. They share DNA ancestry with Turkish cats. It's possible they originally came to Greece aboard fishing boats.

AFRICAN GOLDEN CAT

The shy **AFRICAN GOLDEN CAT** lives in the rainforests of eastern Africa. Their coats can be the color of butter but can also vary from red to gray and black. Not many people have seen one of these wildcats, which are about the size of a beagle. The stocky little felines survive by hunting small monkeys, birds, and sometimes fish.

AMERICAN BOBTAIL

With its naturally short tail, this adorable breed looks like the domestic version of a bobcat. It has the toughness of a hunter but is also known to be affectionate, smart, and supersoft. All **AMERICAN BOBTAILS** are descended from Yodie, who was found as a tabby kitten in Arizona and brought home to Iowa. He later fathered the first litter of kittens that started this popular domestic breed.

AMERICAN CURL

Known for the unique shape of their ears, **AMERICAN CURLS** are cute cuddlers who make an affectionate cooing noise to get the attention of their loved ones. They love to chase toys and explore, then collapse in someone's lap for a nap. These distinctive kitties are born with straight ears that coil during the first few weeks of their lives, then unfold to a gentler curve as they get older.

AMERICAN RINGTAIL

Don't mistake this rare breed of cats for the ringtail cat, a cousin of the raccoon. **AMERICAN RINGTAILS** are known for having tails that form lofty loops. The breed started in the late 1990s after a kindhearted cat lover rescued a tiny, abandoned kitten. She named him Solomon and bred him, hoping that his kittens would share his unique tail.

AMERICAN WIREHAIR

The **AMERICAN WIREHAIR** is very similar to the American shorthair. One hairy difference? These cats have hard, wiry fur. Boxy muzzles give their friendly faces a square shape. They can come in practically any color and pattern. They are not typically very vocal cats, and their meow is quite soft.

ANDEAN CAT

There may only be 1,500 of these beautiful wildcats left in the world, and scientists are working to save them. **ANDEAN CATS** (also called Andean mountain cats) live in the rocky Andes Mountains in Peru and Bolivia, as well as in parts of northern Chile and Argentina. Their plush, ringed tails are nearly as long as their bodies. The main prey of the Andean cat is the mountain viscacha (vis-KA-sha), a rodent related to the chinchilla.

AMERICAN SHORTHAIR

This popular breed is known to be easygoing, curious, and affectionate. They are large, sturdy cats. Adults can weigh between twelve and fifteen pounds. Colors range from striking silver tabbies to beautiful black-and-white tuxedo cats. They have round eyes that are generally some shade of gold or green.

These powerful hunters likely sailed on the *Mayflower*, arriving in America in 1620 with some of the first English settlers. They would have had a job on the big sailing vessels: rodent control. Over the years, these cats continued to be bred to preserve the breed's best traits, so they have a long lineage as tough cats. A lot of **AMERICAN SHORTHAIRS** still work as mousers today, hunting down mice and other pests. The American shorthair was among the first breeds recognized when the Cat Fanciers' Association was established in 1906.

APHRODITE GIANT

These large and friendly felines (also known as Cyprus cats) have a long history of helping humans limit populations of rodents and snakes. The **APHRODITE GIANT** breed likely began with feral (domesticated but living wild) mountain cats on the Greek island of Cyprus. They have long legs (good for climbing tough terrain), big claws, and thick coats. They can weigh as much as twenty-five pounds—which is about the same as an adult male turkey.

ARABIAN MAU

These street-smart Middle Eastern cats developed from hardy, green-eyed desert cats on the Arabian Peninsula. **ARABIAN MAU** cats have short, glossy coats and are highly tolerant to heat. Cats of this breed have unique noses that appear to have a curved bridge, which is the part of the nose between the eyes.

ASIAN SEMI-LONGHAIR

The first **ASIAN SEMI-LONGHAIRS** were the accidental result of a pairing between purebred cats of two other breeds: a Burmese and a chinchilla Persian. These are fluffy, beautiful cats, known to have a jealous streak when other pets get attention. They are also known as the Tiffanie cats.

ASIATIC GOLDEN CAT

An **ASIATIC GOLDEN CAT'S** pelage (a mammal's hair, wool, or fur) is usually light red. Asiatic and African golden cats look very similar, but they are actually not closely related. Asiatic goldens carry their tails with a tidy curl at the tip. They come out at night in the forests of Nepal and northeast India as well as China, Thailand, Malaysia, and Sumatra. In China they are called rock cats. Sometimes they share their habitat with giant pandas.

AUSTRALIAN MIST

This burly breed is the first pedigreed cat of Australia and started as a cross between two ancient breeds: the Abyssinian and the Burmese. **AUSTRALIAN MISTS** are lively kitties whose coats often have a marbled tabby pattern. They come in seven colors: brown, blue, chocolate, lilac, caramel, gold, and peach. Their ticked (which means "lightly spotted") short fur gives them the soft, misty appearance that inspired their name.

BALINESE

These beautiful blue-eyed cats have the same stunning black or brown points as their ancestors: Siamese cats. **BALINESE** cats were originally bred in the United States when Siamese cats gave birth to litters of plumy-tailed, silky-haired kittens. Even though the Balinese became its own breed, they are still often called long-haired Siamese cats.

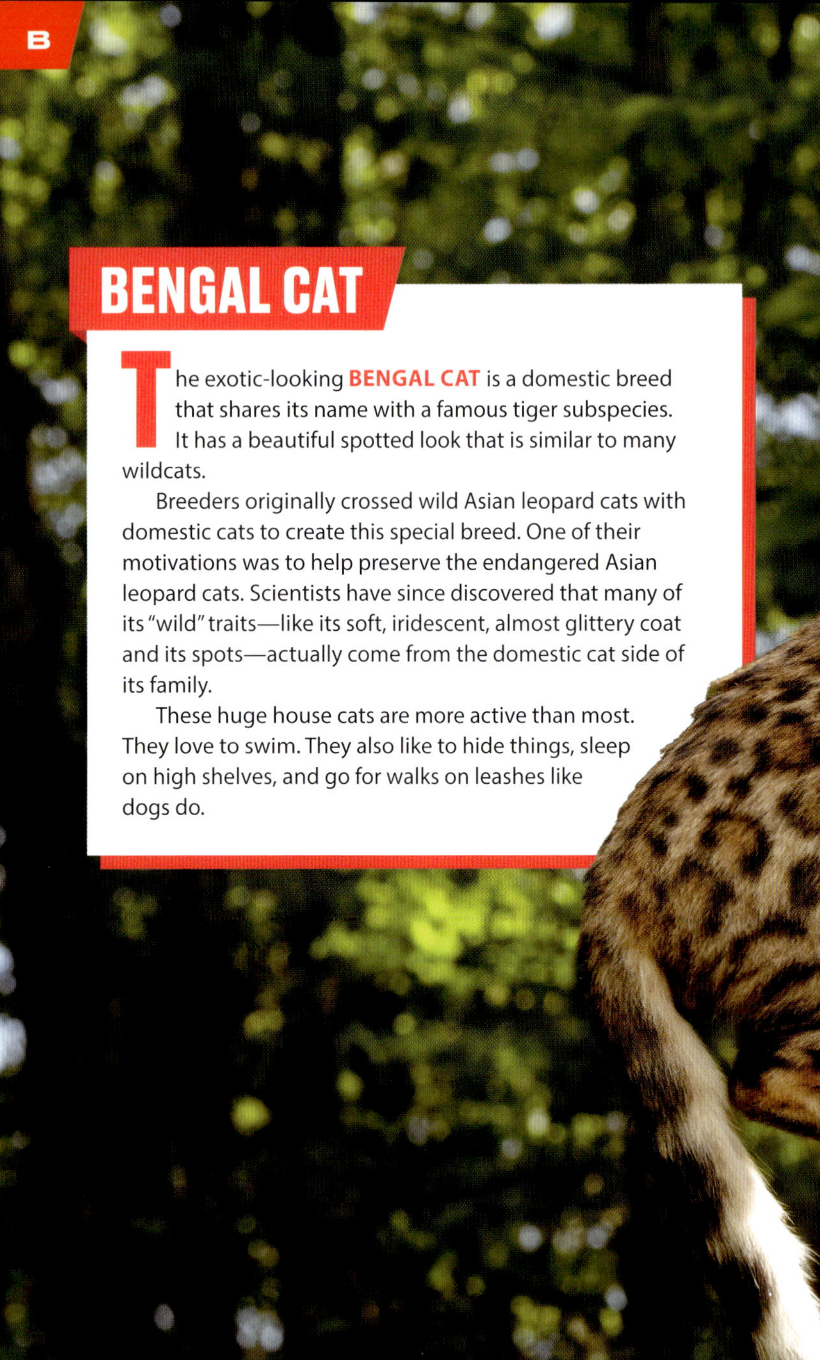

BENGAL CAT

The exotic-looking **BENGAL CAT** is a domestic breed that shares its name with a famous tiger subspecies. It has a beautiful spotted look that is similar to many wildcats.

Breeders originally crossed wild Asian leopard cats with domestic cats to create this special breed. One of their motivations was to help preserve the endangered Asian leopard cats. Scientists have since discovered that many of its "wild" traits—like its soft, iridescent, almost glittery coat and its spots—actually come from the domestic cat side of its family.

These huge house cats are more active than most. They love to swim. They also like to hide things, sleep on high shelves, and go for walks on leashes like dogs do.

BIRMAN

These cloudlike kitties have a distinctive coat color called mink. The **BIRMAN** is thought to have originated from ancient cats who guarded temples for an ancient civilization of people called the Khmer, who lived in Burma (now called Myanmar). The Khmer believed the cats could keep away evil. Birman cats have piercing sapphire-blue eyes and white feet (called gloves).

BLACK-FOOTED CAT

This feisty feline is the smallest wildcat in Africa. Round, low-set ears help it hear the rodents and birds on which it preys. One little **BLACK-FOOTED CAT** can eat three thousand rodents in a year. These cats also like to eat eggs. They have hair on the black soles of their feet to prevent them from being burned on the hot sands of southern Africa. Wild neighbors include the African wildcat, Cape fox, and black-backed jackals.

BOBCAT

Known for the black tufts of hair on its ears and its short—"bobbed"—tail, the **BOBCAT** is the smallest type of lynx. Bobcats are fierce predators and can even kill small deer, but they're only about the size of a cocker spaniel—a medium-size dog. They live in North America. Occasionally they hunt porcupines, flipping the prey over to eat the soft, quill-free underside. They are known to make a noise that sounds like a screaming woman.

BOMBAY

The **BOMBAY** is no ordinary black cat. This exotic feline was originally bred by crossing black American shorthairs and Burmese cats. The result? A sweet, chatty, miniature black panther that is a lot cuddlier than the wildcat it resembles. Bombay cats are known to be social, bossy snugglers who prefer to be under the covers with their favorite humans. The breed's name refers to the large Indian city Bombay (now known as Mumbai), where true black panthers once roamed wild.

BORNEO BAY CAT

The rarely seen **BORNEO BAY CAT** is about the same size as a house cat, with a long, white-tipped tail and huge eyes. These highly endangered wild felines are found only in the forests of Borneo, an island in the southwestern Pacific Ocean. Unfortunately, they are disappearing along with their forest home as humans destroy their habitat.

BRAZILIAN SHORTHAIR

The **BRAZILIAN SHORTHAIR** is a breed that started with feral street cats. Some say they arrived in Brazil in the 1500s on ships alongside Portuguese sailors. These energetic kitties are known for their playful and mischievous antics, as well as their intelligence. They are similar to American and British shorthairs but are slightly more muscular.

BRITISH LONGHAIR

With hair in patterns from tabby to tuxedo, the **BRITISH LONGHAIR** cat is a close relative of the British shorthair. This breed was developed after World War II, when the British shorthair had nearly been wiped out. Breeders restored the population of these cats by breeding American shorthairs with other purebreds such as Persian. Any long- or medium-haired kittens in the resulting litters didn't qualify as British shorthairs, so they became a beautiful new breed.

BRITISH SHORTHAIR

The plucky **BRITISH SHORTHAIR** has been a favorite in England, Scotland, and Wales for centuries. They likely were brought by the Romans (who conquered the island in AD 43 and ruled there for almost four hundred years), becoming popular companions thanks to their work on farms. Many British shorthair cats also worked on ships to hunt rodents that snuck aboard. They are known for their compact and muscular bodies, plump cheeks, plush coats, and big, round eyes.

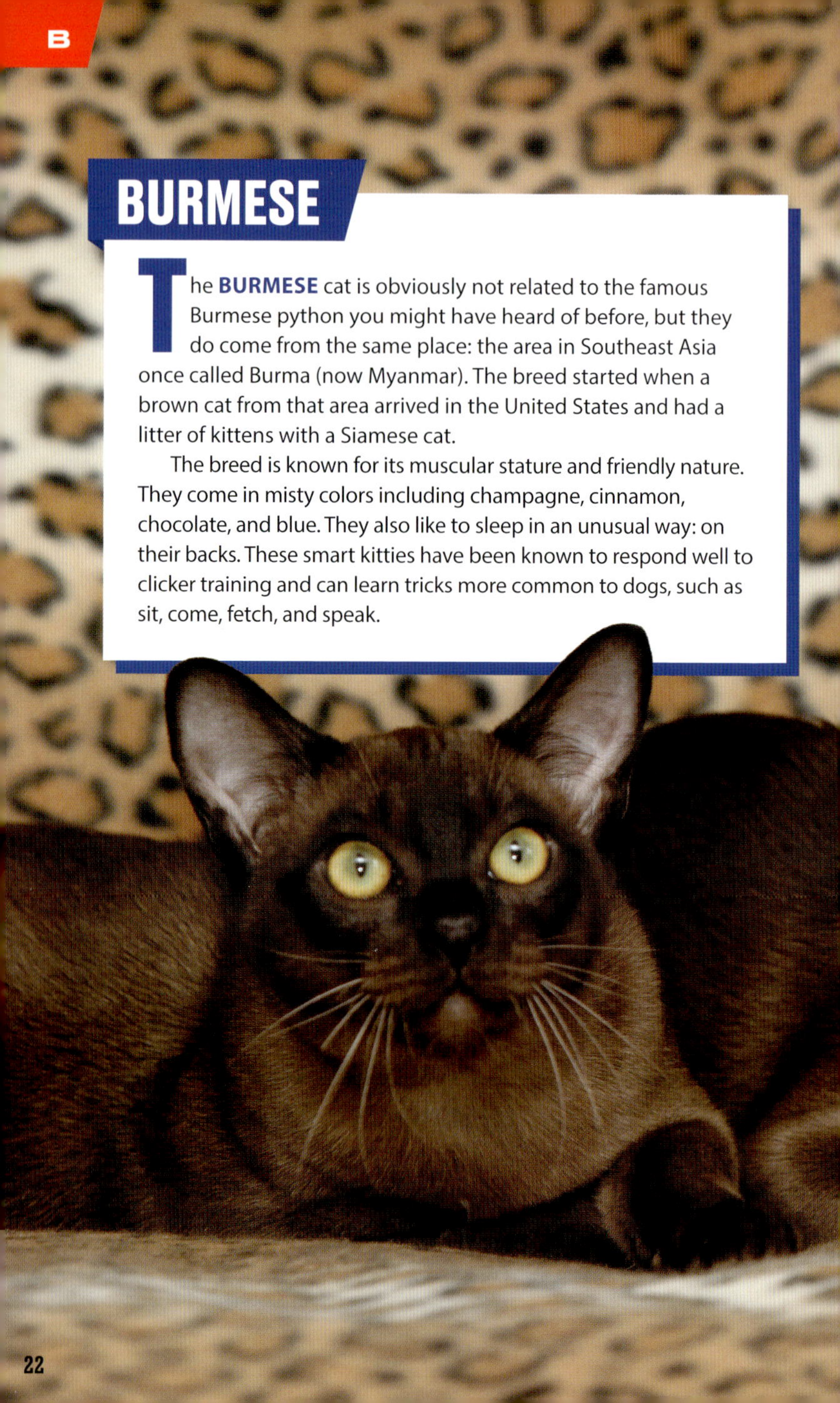

BURMESE

The **BURMESE** cat is obviously not related to the famous Burmese python you might have heard of before, but they do come from the same place: the area in Southeast Asia once called Burma (now Myanmar). The breed started when a brown cat from that area arrived in the United States and had a litter of kittens with a Siamese cat.

The breed is known for its muscular stature and friendly nature. They come in misty colors including champagne, cinnamon, chocolate, and blue. They also like to sleep in an unusual way: on their backs. These smart kitties have been known to respond well to clicker training and can learn tricks more common to dogs, such as sit, come, fetch, and speak.

CALIFORNIA SPANGLED

Members of this long, lean, rare breed might look like little leopards, but their personalities are pure house cat. **CALIFORNIA SPANGLED** cats love to cuddle and pounce on toys. These exotic kitties are an American invention, first bred in California in the 1980s from a mix of five cat breeds: Abyssinian, Siamese, Angora, American shorthair, and British shorthair.

CANADA LYNX

The **CANADA LYNX** is a leggy, large relative of the bobcat. A fluffy ruff rings its face. Paws that spread out like snowshoes help this hungry hunter travel through deep snow to find its prey. What's on the menu? Snowshoe hares, mostly. This prey is a species similar to rabbits. But a Canada lynx might also snack on birds and even take down larger prey such as a young caribou, which is a reindeer.

CARACAL

A hungry **CARACAL** (ker-UH-kal) will leap into the air and claw a bird in flight for a quick meal. These wildcats are known to prey on lagomorphs (rabbits and hares). They also hunt larger species, such as antelopes and ostriches. Their enormous ears, tipped with long tufts of black fur, swivel to home in on a target. They can be found across Africa and western Asia.

CHANTILLY-TIFFANY

The regal **CHANTILLY-TIFFANY** has a fluffy, plumed tail and big ear tufts with wisps called streamers. They come in colors including blue, fawn, cinnamon, black, and lilac. Originally, these cats were thought to be semi-longhair Burmese cats. This is because the original Chantilly-Tiffany litter came from two brown cats whose breeds were unknown. It was originally called the foreign longhair and is a completely separate breed from the Tiffanie cat.

CHARTREUX

Chartreux cats always seem to smile. The breed was developed long ago in France and hasn't changed much over the centuries. Some would describe this elegant, powerful breed's stout physique as semicobby, meaning that they have short legs and a compact body. **CHARTREUX** cats have distinctive dense blue-gray fur, orange eyes, gray noses, blue lips, and rose-colored toes.

CHAUSIE

The name of the **CHAUSIE** (CHOW-see) breed comes from the scientific name of the wild jungle cat, *Felis chaus*, found in the area of the Caspian Sea and South Asia. The domestic chausie is built a lot like its wild namesake: athletic, long legged, and well balanced. They are often said to look like miniature mountain lions.

CHINESE MOUNTAIN CAT

This small, stocky cat is found only in China's mountain steppes and alpine meadows, where it blends into its grassland habitat. To some locals, the **CHINESE MOUNTAIN CAT** is known as the grass cat. This wild feline looks like a miniature mountain lion. Fur between its toes helps keep it warm during harsh winters. Piercing blue eyes help it hunt voles and other rodents.

COLOCOLO

This small South American wild feline varies in coat color and pattern depending on where it lives—from high in the Andes Mountains to lower elevations in Argentina and Brazil. In colder areas, they have thicker fur so they can stay warm. They like to eat small rodents such as leaf-eared mice. **COLOCOLOS** are also known to get in trouble with humans for snatching up their domestic goats and chickens. Their name refers to an ancient warrior chief. These cats are also known as the Pampas cat—the "Pampas" are an area of grasslands in South America.

CHEETAH

Cheetahs are famous for their speed and their spots. These lanky cats are the fastest land animals on earth, able to sprint up to seventy-five miles per hour—faster than a racehorse. Their black markings likely help them blend into the scenery for safety from bigger cats and for hunting. They can be found in Africa and northern Iran. Their favorite habitats? Grasslands, forests, and arid scrublands.

Charming **CHEETAHS** purr, churr, and chirp in friendly communications with each other. They also hiss and growl like house cats if threatened. One thing they can't do? Roar like other big cats.

As much as humans admire this spectacular species, cheetahs are in grave danger of going extinct due to habitat loss and poaching (illegally hunting animals for their skins and other body parts). As a result, fewer than six thousand cheetahs remain in the wild today.

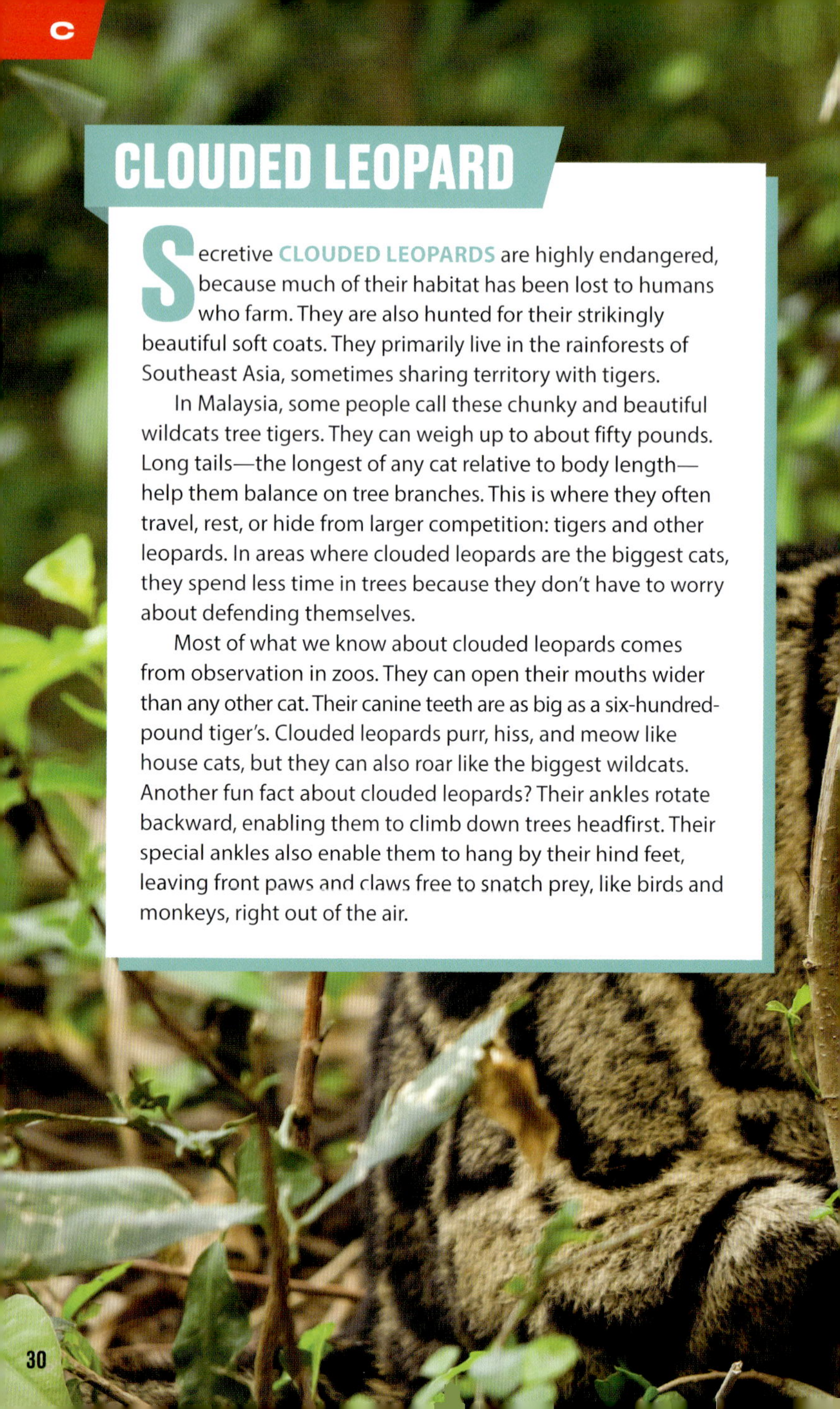

CLOUDED LEOPARD

Secretive **CLOUDED LEOPARDS** are highly endangered, because much of their habitat has been lost to humans who farm. They are also hunted for their strikingly beautiful soft coats. They primarily live in the rainforests of Southeast Asia, sometimes sharing territory with tigers.

In Malaysia, some people call these chunky and beautiful wildcats tree tigers. They can weigh up to about fifty pounds. Long tails—the longest of any cat relative to body length—help them balance on tree branches. This is where they often travel, rest, or hide from larger competition: tigers and other leopards. In areas where clouded leopards are the biggest cats, they spend less time in trees because they don't have to worry about defending themselves.

Most of what we know about clouded leopards comes from observation in zoos. They can open their mouths wider than any other cat. Their canine teeth are as big as a six-hundred-pound tiger's. Clouded leopards purr, hiss, and meow like house cats, but they can also roar like the biggest wildcats. Another fun fact about clouded leopards? Their ankles rotate backward, enabling them to climb down trees headfirst. Their special ankles also enable them to hang by their hind feet, leaving front paws and claws free to snatch prey, like birds and monkeys, right out of the air.

COLORPOINT SHORTHAIR

This outgoing, elegant designer breed originally started when a red American shorthair and a seal-brown Siamese cat had a litter of kittens. The idea was to breed a Siamese cat that would end up with red points—the darker fur on its face, feet, ears, and tail. **COLORPOINT SHORTHAIRS** have a unique coat that comes in beautiful colors including cream and red. They are also known to be very loud, capable of making more than one hundred vocal sounds.

CORNISH REX

Cats of this unusual breed are affectionate and social, often sitting in a warm lap. **CORNISH REX** cats have long faces, bat-like ears, and legs like a greyhound dog. They're also known for their extremely short, soft coat that comes in a range of colors and patterns from tabby to tuxedo and smoky gray to tortoiseshell.

DEVON REX

Both the **DEVON REX** and the Cornish rex breeds originated in the United Kingdom. Both have lean bodies and big ears. The laid-back Devon rex has lower-set ears and shorter whiskers. It also has a wavy coat and a fun-loving nature.

DOMESTIC LONGHAIR

Domestic longhair cats are the same as domestic shorthairs, with one obvious exception: They have medium or long plush hair. They are not a specific breed, but a loveable and popular mixed breed. **DOMESTIC LONGHAIR** cats can come in a vast range of different colors, patterns, and sizes thanks to the many different "mixes" they can come in.

DOMESTIC SHORTHAIR

Many treasured rescue cats are considered **DOMESTIC SHORTHAIRS**. They are similar in stature to American shorthairs, but they are not members of an established breed. Domestic shorthairs are marvelous mixed breed cats with—you guessed it: short hair. Domestic shorthair cats make up ninety-five percent of the cat population in the United States, which is why this kind of cat is frequently seen at animal shelters around the country.

DONSKOY

This remarkable breed is also known as the Russian hairless cat. It's known to be a cuddler—especially when it's cold out. They are curious and mischievous. The start of the breed was a partially hairless cat named Varvara who was rescued in 1986. **DONSKOY** cats have beautiful almond-shaped eyes and long toes.

DRAGON LI

This fiery ancient breed likely originated from the wild Chinese mountain cat. **DRAGON LI** cats are also known as li hua (lee WAH), which means "fox flower." They often have a tabby pattern with the unmistakable *M* on their foreheads. Black dots at the corners of their mouths make them look like they're always smiling.

It's rare to find one of these gentle cats with striking eye markings outside of China, where it is a treasured part of folklore, art, and culture. The dragon li cat is known for its charming personality, smarts, loving and playful nature, and tendency to make a lot of noise. As they have with many other cats, humans have relied on dragon li cats throughout history to keep populations of rodents down; they tend to be powerful rat hunters.

EGYPTIAN MAU

Some elegant breeds of cats, including the **EGYPTIAN MAU**, were depicted in ancient Egyptian art and culture. In fact, scientists have been able to confirm that today's cats are related to the cats of ancient Egypt by comparing the DNA of living cats to the DNA of mummified cats found in tombs. Known for their bright green eyes and coats in silver, smoke, or bronze with tabby markings, these unique felines love to sing.

EURASIAN LYNX

Two hundred years ago, so few **EURASIAN LYNX** existed that scientists considered them nearly extinct. Over the last fifty years, conservationists have reintroduced this wild predator to countries including Switzerland, Croatia, and France. There are now thought to be more than seventeen thousand of these cats in Europe. They hunt by hiding in shrubs to ambush deer, squirrels, mice, birds, and hares. These big cats share their forest home with wolves.

EUROPEAN SHORTHAIR

The **EUROPEAN SHORTHAIR** developed as a breed hundreds of years ago without much human intervention. These adventurous, shy cats lived mostly in the northeastern part of the continent. Today, most European shorthair breeders are in Scandinavian countries.

EUROPEAN WILDCAT

These fluffy felines might be mistaken for domestic tabbies, except for their especially fluffy raccoon-like tails. They live in Europe, Russia, and Asia. **EUROPEAN WILDCATS** are excellent climbers, but hunt mostly on the ground. They move slowly through their territory, waiting quietly to pounce on small prey. These fearless wildcats can live to be fifteen years old.

EXOTIC SHORTHAIR

Exotic shorthairs are the short-haired cousins of Persian cats. Both breeds are easily recognizable with their flat, broad noses that give their faces a squished look. **EXOTIC SHORTHAIRS** have a plush coat that's so thick and soft it feels a bit like a teddy bear.

FISHING CAT

The powerful, pink-nosed **FISHING CAT** is a great swimmer. Its front toes, partially webbed, help it catch fish. It also hunts at the edge of creeks and marshes, scooping up shellfish and amphibians. It'll eat anything it can get its paws on, such as small mammals, birds, and lizards.

Fishing cats can be found in wetland areas across Pakistan, India, Cambodia, and farther north into the Himalayan foothills. Dense strands of hair keep them warm in cold water. Very few humans have seen them in the wild. Camera traps help scientists monitor their presence in protected areas.

FUN FACT: The fishing cat is very different from another predator with a similar name, the fisher, which is a kind of weasel.

FLAT-HEADED CAT

This small feline predator has a flatter forehead than most other cats. It is one of the most endangered cats in the world. The **FLAT-HEADED CAT** is built for water and considered semiaquatic, similar to a beaver. Its long, narrow jaws with backward-facing teeth help it grab and hold on to slippery fish and frogs. These cats live in the wetlands of Sumatra and Borneo, and on the Malay Peninsula. Like many cats, they are crepuscular, which means they are most active around twilight.

GEOFFROY'S CAT

These petite predators weigh between six and twelve pounds. The species is named after a French zoologist, Étienne Geoffroy Saint-Hilaire. Despite their small size, these rugged wildcats live in some of South America's harshest terrain. **GEOFFROY'S CATS** survive cold winds, dust storms, and extremely hot temperatures. They have a waterproof coat like a seal's, which helps keep them dry and insulates them from extreme heat or cold.

GÜIÑA

Don't mistake this cute kitty for a cheetah cub. Full grown, this wild predator is about half the size of a house cat. The **GÜIÑA** (GWEE-nyuh) chirps like a bird when excited. It has rounded ears like a ferret, huge eyes, and a big bushy tail. The güiña is also known as the kodkod, or the Chilean cat, because it lives in the rainforests of Chile and western Argentina. These kitties will eat whatever they can catch—from mice and birds to anything else small that moves, including bugs.

HIGHLANDER

The **HIGHLANDER** cat is a cuddling wonder from down under. This Australian breed is playful, curious, and fun-loving. Some of these cats have adorably curled ears and bobbed tails. Their large size makes them a "giant" breed—adult males can weigh up to twenty pounds!

HIMALAYAN

In the 1930s, breeders of Siamese and Persian cats began crossing them together to better understand how they passed colors and patterns onto kittens. In the process, they created a completely new breed: the sweet **HIMALAYAN** cat. "Himmies" are mild-mannered, affectionate, and sometimes shy cats. Some special Himalayan cats have worked as Hollywood actors and starred in features like the Homeward Bound movies and *Meet the Parents*.

IBERIAN LYNX

The Iberian Peninsula is a mountainous region surrounded by water on three sides, mostly made up of Spain and Portugal. It's also home to the long-legged **IBERIAN LYNX**. This large wildcat, which sports tufted ears and a bearded chin, is at risk. In 2002, there were fewer than one hundred left in the wild, thanks to diminishing sources of food (mainly wild rabbits) and human impacts such as cars, destroyed habitats, and hunting. Today conservation efforts are helping—the population of Iberian lynx living wild on the peninsula has more than quadrupled. Each one of them is helping to save their special species.

JAGUAR

The **JAGUAR** is a big cat with a big name. It comes from yaguara—in the language of an indigenous Brazilian people, the Tupi-Guarani—which roughly translates to "he who kills with one leap." This cat has many claims to fame: It is the largest cat in the Western Hemisphere (roaming across Latin America). Jaguars are the third largest of all the big cats, after the tiger and the African lion. Adult male jaguars can weigh as much as 265 pounds.

While most species of cats avoid water, jaguars regularly take the plunge. They dive into murky water to hunt prey such as the semiaquatic rodents called capybaras, or glide from tree to tree to sneak up on monkeys and sloths. Scientists have also reported that these cats attract fish by waving their tails over the water.

Jaguars are often confused with a similar species, leopards, which live in Africa and Asia. You can tell the difference between the two big cats by zooming in on a photo of their spots. A jaguar's spots are blocky rosettes, or circular markings, with distinct dark dots inside. Leopards also have rosettes, but without the dot inside. Their size also provides clues: A jaguar has a chunkier build with shorter and very strong legs—perfect for pouncing and leaping.

JAGUARUNDI

The **JAGUARUNDI'S** eyes seem to smolder. Their long, slender bodies and short-haired coats make them look a bit like otters. The Jaguarundi's ears are all one solid color, from the backs of them to the insides. This gives them a unique appearance compared to many other cats. This breed is a close relative of the cheetah and the mountain lion. One remarkable thing about this cat? It is especially chatty, and uses more than ten noises to communicate, from purrs and whistles to chatters and chirps.

JAPANESE BOBTAIL

Elegant and spunky, these royal housecats (with their signature naturally short tails) were Japanese aristocracy in ancient times. In 1602, however, the Japanese government called on these pampered pets to help fight off the invasion of mice and rats that threatened the country's precious silkworms. These silkworms spin cocoons that are used to make silk. Households released their pampered pets into the streets to hunt, helping to save the country's silk industry.

JUNGLE CAT

The tawny **JUNGLE CAT** lives in Africa and Asia. Also known as the swamp cat or reed cat, it prefers a habitat among the tall grasses that surround wetlands. Why? Because this is also a prime rodent residence. One of these cats can eat as many as five rodents a day. Remains of revered jungle cats, like the remains of a few other species of cats, have been found mummified in ancient Egyptian tombs.

KHAO MANEE

The **KHAO MANEE** is a striking, delicate breed. They originated centuries ago in Thailand, where they're considered sacred animals. These glamorous kitties have pure white coats; this breed's name translates to "white gem" in Thai. They also have eyes that glitter like gemstones and aren't always the same color. When a cat's—or any creature's—eyes are two different colors, it is called heterochromia.

KORAT

Considered to be a good luck charm in Thailand, the **KORAT** is a stunning blue-gray. For centuries these ancient, sacred cats were not sold but instead were gifts given as tokens of respect and good fortune. Locals refer to the breed as si sawat, which means "the color of sawat seed" in Thai. Sawat seeds are grayish green, so this name is certainly fitting. Their toe beans (a nickname for a cat's paw pads) come in special colors, including dark blue and lavender with a pink tinge.

LAPERM

The first **LAPERM** cat was born in the 1980s in an Oregon barn. It came out completely bald, then developed a soft, curly coat of hair as it grew up. That cat went on to bear many more curly kitties, and thus the LaPerm breed was born. The word *perm* refers to a process for curling hair that was popular during that decade. These cats' curls range from slight waves to full-on ringlets and lovely long-haired corkscrew curls.

LYKOI

The **LYKOI** is a cat that looks like a wolf. These warm, loving, and devoted lap cats are also called wolf cats or werewolf cats. Their distinctive coat—wiry black fur tipped with white on pink skin—has shown up in feral cats for decades. One breeder matched unrelated cats with this look in 2011 and bred a litter of adorably fuzzy—and slightly odd—werewolf kittens. The breed's name comes from the Greek word for wolf: *lykos*.

LEOPARD

On a safari in Africa, you might search for **LEOPARDS** for days without seeing one. They are secretive and stunning. They are also among the most endangered cats in the world. They live in many habitats, from the desert to mountains and rainforests, across sixty-two countries in Africa and Asia. Their appetite varies from big antelope such as kudu to smaller snacks including fish, rodents, and even dung beetles. Scientists and conservationists use technology such as camera traps to monitor populations to help make sure this superlative species—the smallest of the five big cat species—doesn't go extinct. The information they collect helps them educate people and create regulations to protect leopards.

LION

Imagine sitting in a safari vehicle in the dark. The moon lights up South Africa at night. Then you hear it—the sound is similar to a rumble in your chest. The rumble turns to a roar that makes your hair stand on end. **LIONS!**

Seeing (or even just hearing) a lion in the wild is rare and wonderful. Lions live in the grasslands, forests, and desert areas of sub-Saharan Africa. Experts report that there are about twenty-four thousand left in the wild. A hundred years ago, there were more than two hundred thousand roaming free. The biggest threats to lions are humans, who hunt them illegally and destroy their habitats for the development of buildings and agriculture.

Lions have some remarkable characteristics. Magnificent manes on male lions attract females and make them look more intimidating to competitors. Huge teeth and claws help them bring down large prey such as buffalos, giraffes, and zebras. Most vertebrates, or creatures that have bony skeletons, have a bone at the base of their tongue called the hyoid, which helps to control vocalization, swallowing, and similar actions. Lions (and the other cats that roar—tigers, leopards, and jaguars) have a more flexible hyoid. This enables the lion to make its unforgettable roar, which can be heard from five miles away.

MAINE COON

What cuddles like a lap cat but has the face of a lion and the playfulness of a golden retriever? The **MAINE COON** cat! This spectacular old breed originated in the United States in Maine. It likely resulted from pairing local domestic shorthairs with long-haired cats such as Angoras.

These cats are hardy and handsome. They are also well-adapted to harsh winters. Their glossy fur repels water. The fluffy britches on its belly and rear protect it from wet snow. A long, bushy tail helps keep it warm. Tufted feet work like insulated snowshoes. Even its large ears are lined with fur. Another thing that sets this cat apart from other breeds? Its large size. Adults usually weigh twelve to eighteen pounds. Some are even larger—a famous Maine coon named Kefir weighs twenty-eight pounds and, when he stands on his hind legs, is taller than a four-year-old child!

MAINLAND LEOPARD CAT

The versatile **MAINLAND LEOPARD CAT** slinks through the forest as easily as it swims in a river to hunt frogs, birds, squirrels, rats, insects, and lizards. It's the only wildcat left in Singapore since the extinction of tigers and leopards from the Southeast Asian island country, but its habitat extends through large areas of mainland Southeast and South Asia. This wild feline is only about the size of a house cat, and it has a unique stumpy tail that's only half as long as its body.

MANX

This playful and social breed originated on the beautiful Isle of Man in the Irish Sea. The breed has appeared in stories and tales from as far back as the time of the Vikings. Local legend says that a pair of these cats were the last to board Noah's ark—and in the rush, their fluffy tales got lopped off by the closing door. A single **MANX** litter can include kittens with a regular tail, with a stubby tail, or with no tail at all. Their long back legs give them an unusual way of moving that looks similar to a bunny hopping. These cats are very social and often follow their owners around, always wanting to be near them.

MARBLED CAT

The **MARBLED CAT** lives in the forests of South and Southeast Asia. Its large feet and its tail help it balance on branches high in the trees. The forest canopies provide a handy buffet for the marbled cat. It eats birds, squirrels, and reptiles.

MARGAY

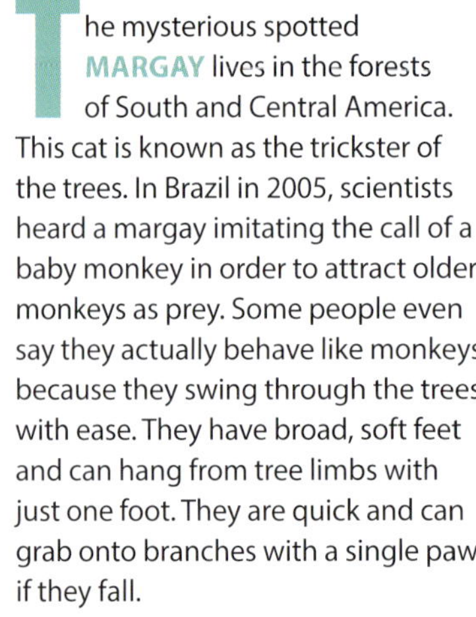

The mysterious spotted **MARGAY** lives in the forests of South and Central America. This cat is known as the trickster of the trees. In Brazil in 2005, scientists heard a margay imitating the call of a baby monkey in order to attract older monkeys as prey. Some people even say they actually behave like monkeys because they swing through the trees with ease. They have broad, soft feet and can hang from tree limbs with just one foot. They are quick and can grab onto branches with a single paw if they fall.

NEBELUNG

The **NEBELUNG** has a misty, mythical appearance. Its name is rooted in the German word *nebel*, which means fog. This breed was created in the 1980s and 1990s to resemble Russian cats from the nineteenth century. The first cats registered to this plush, silver-tipped breed were named Siegfried and Brunhilde, two German names. Even so, these kitties do not have German origins. They first appeared in the United States when a woman gave her son a black domestic shorthair cat and it then had a litter of kittens fathered by a Russian blue cat. Thus, the Nebelung cat was created. These green-eyed felines tend to be playful and friendly. They love their families but might be shy with strangers.

NORWEGIAN FOREST CAT

Norwegian forest cats are known to be strong, sensitive, beautiful, and intelligent. The "Wegie" (WEE gee) is a mild-mannered cat that can take up to five years to grow into adulthood.

Their lavish coats have had an important purpose throughout the breed's history: to protect it from the often cold, wet Scandinavian forests where it first originated. **NORWEGIAN FOREST CATS** shed their heavy coats twice a year. The term *britches* refers to the fluffy fur on the back of their rear legs.

These beloved cats (who were also good at hunting rats on ships and anywhere else they lived) have been referred to in Viking legends as skogkatt, a Norwegian word that roughly translates to "forest cat." In one tale, the Norse goddess Freya rides a chariot pulled by six of these giant, impressive cats. This cold-weather kitty is also called Hulderkatt or "troll cat" in some Norwegian folktales.

OCELOT

The **OCELOT** stalks, crouches, and then, like a strike of lightning, pounces on a raccoon. The unsuspecting prey never saw the fierce predator hidden in the tree canopy. This small, spotted golden cat lives in several different areas, from the southernmost parts of Texas to South America. In Florida, there is a small wild colony that may have originated when humans released some ocelots they kept illegally as pets.

An ocelot's body is about three feet long, not including the tail. It hunts rodents, small mammals, iguanas, deer, and wild pigs called peccaries. But these cats are also prey. Mountain lions, jaguars, anacondas, and raptors—such as huge harpy eagles—hunt ocelots.

Humans nearly hunted these beautiful cats to extinction for their fur. A single fur coat required pelts from twenty-five ocelots. Today in the United States and many other countries, it's illegal to hunt an ocelot or sell its fur. Still, this species struggles to survive as its forest home is cleared to make room for farming, development, and roads.

PALLAS'S CAT

The **PALLAS'S CAT** might look like a slightly grumpy and super-fluffy house cat, but it's actually a brilliant hunter. To survive in its often-snowy home in Central Asia, it must catch and eat a steady diet of voles and sometimes other rodents, lizards, and an occasional hare, each day. It has also been known to prey on small wild sheep. This feline species tends to hunt at night—to avoid being hunted itself by a fox or an eagle. From some angles, the Pallas's cat (on its adorably short legs) appears wider than it is tall.

PERSIAN

Poofy. Plush. Posh. **PERSIAN** cats have the most luxurious long coats. This ancient breed is the parent breed of many other famous types of cats, including the Himalayan. The Persian breed is believed to have started with long-haired cats in Italy in the 1600s through cats who had been brought there from Persia. Scientists have discovered this breed within hieroglyphics from as far back as 1684 BC. Many centuries later, Queen Victoria is said to have had some Persian kitties of her own, which she adored. Who could blame her?

P

PUMA

Mountain lion. Cougar. Catamount. Panther. Ghost cat. These are just some of the names for this formidable and beautiful big cat. This wild and mysterious predator lives throughout North and South America. In some areas, they coexist with other top predators, including wolves and grizzly bears.

Powerful legs can launch the **PUMA** into leaps forty-five feet long and fifteen feet high. They can sprint up to fifty miles per hour. A puma will often cache, or save, leftovers of prey like deer and rabbit by burying it. In the mountains of Chile, they survive by hunting a relative of the llama called the guanaco.

Researchers use camera traps to learn as much as they can about this very secretive species. In one study, scientists installed cameras in wild dens to observe how mother pumas interact with their young. They discovered that mother cats purr to their kittens to communicate. They stay with their babies day and night until the kittens are about ten days old. Then they teach their young to hunt, first by letting them observe.

RAGDOLL

The **RAGDOLL** may be the floppiest breed in existence. That's how it got its name. These kitties are so easygoing that they tend to go limp when snuggled, cuddled, or held. They have beautiful blue eyes and long, fluffy hair. Adults can grow to be twenty pounds. These social butterflies are outgoing and friendly, making them popular pets.

One famous ragdoll? A rescue kitty named Benjamin Button owned by superstar Taylor Swift. Benjamin, an adorable übercuddler, has appeared on multiple magazine covers with his cat-loving owner. She met him when he worked as a kitty actor on the set of one of her music videos in 2019. He was part of a program to help publicize the plight of homeless cats. Swift fell in love with the kitten and adopted him. Benjamin has two feline sisters, both of whom are Scottish folds and are named after TV characters: Meredith Grey and Olivia Benson.

RESCUE CATS

A plump, brown, short-haired cat named Mochi who loves to snuggle and relax inside her kitty tent. A fluffy long-haired beauty named Moonshine who loves car rides with his family to spend time at their vacation home. A spotted kitten named Cedric Mullins who sneaks snacks from his doting owner's pantry. A green-eyed tortoiseshell cat named Bella who loves to boss around her three large canine companions and oversaw the writing of this book. These beloved felines are all different, but they all have something in common: No matter their shape, size, color, or ancestry, they were once homeless.

Thousands of rescue groups and countless volunteers work tirelessly to find forever homes and happy endings for the more than three million homeless cats in this country every year. **RESCUE CATS** range from newborns being bottle-fed around the clock to sweet seniors who just need a warm place to lay their heads and someone to love them. Some rescue organizations have opportunities for kids and their families to foster homeless cats and kittens to help keep them out of the dreary cages they'd otherwise be stuck in. If your family wants to get a cat or kitten, consider adopting before buying. There are even rescue groups for specific breeds and types of cats you might have learned about in this book.

RAAS

Have you ever heard of the **RAAS** cat? Most people haven't. It's a rare and exotic (and only recently domesticated) breed from Raas, a sunny, beachy Indonesian island. This lanky breed of cats is endemic to the island, which means it only occurs naturally in that geographic area. It's uncommon to find this breed anywhere else.

RUSSIAN BLUE

The **RUSSIAN BLUE** has a delicate build and a short coat that only ever comes in one color: deep gray with silver tips that seem to glitter when the cat moves. Its eyes are usually bright green. These cats originated in northwest Russia and are rumored to have once been the treasured cats of Russian czars.

RUSTY-SPOTTED CAT

The most amazing thing about the **RUSTY-SPOTTED CAT** is not its spots! This adorable wildcat, found in India, Sri Lanka, and Nepal, is the world's smallest cat species. Full grown, one would fit in the palm of your hand. Some call this species the hummingbird of cats. But don't measure its boldness by its kittenish size. The fierce and agile rusty-spotted cat sees through the forest with eyes that are six times as sensitive as a human's. It hunts birds, frogs, and sometimes even small mammals that weigh more than it does.

SAND CAT

It's easy to miss the **SAND CAT** in its native desert home. Its pale coat blends into the scenery around it. Sand cats live in Africa's sweltering Sahara desert, the hot Arabian Peninsula, and arid areas of central Asia. They can go without water for weeks and instead stay hydrated with the blood of their prey: snakes, rodents, birds, and spiders.

SAVANNA TIGER CAT

This special species of wildcat has also been referred to as the oncilla, the tigrina, and the small-spotted cat. **SAVANNA TIGER CATS** are rather dainty. Males weigh about eight pounds at most. Members of this species have surprisingly large territories in Costa Rica, Panama, and central Brazil. When threatened, the savanna tiger cat will arch its back, show its teeth, and make a wild whistle to scare danger away.

SCOTTISH FOLD

This adorable breed is most recognizable for its lop ears and cute face. Reports of cats with folded ears go back as far as the 1700s. The **SCOTTISH FOLD** breed started in the 1960s in Perthshire, Scotland, with a farm cat with droopy ears named Susie. Today, folds are more popular in the United States, but they can be difficult to breed because of the genetic mutation which causes the folded ears: It is best for one parent to have straight ears and one parent to have folded ears. This prevents joint disease.

SELKIRK REX

Is it a stuffed animal or a cat dressed up like a sheep? The **SELKIRK REX** is a cuddly, curly cat. Even its whiskers curl. Selkirk rex cats tend to be mellow and calm. They enjoy sitting in laps. The very first kitten of this breed was named Miss DiPesto because her fur was similar to that of a curly-haired character named Miss DiPesto in the television show *Moonlighting*. The breed originated in Montana.

SERVAL

A serval struts on superlong legs through the grass on the prowl for supper. Huge ears swivel like satellites, listening for the sounds of small mammals, reptiles, and amphibians. The **SERVAL** might leap six feet in the air to snag a bird while it's still in flight. It might slosh into a river to snatch a slippery frog (a single serval can catch thirty frogs in a single afternoon). It might pounce on an unsuspecting rat, or even a grasshopper. These impressive hunters live all over Africa in a variety of landscapes.

What is one thing a serval has plenty of? Spots. Different spots on a serval's body serve different purposes. Spots on their coats help them blend in with the dappled light of trees. Some spots merge into stripes that provide camouflage in tall grass. Spots on the back of a mother cat's ears help her cubs keep track of her and identify her.

Servals can live an average of ten years in the wild. However, they face a lot of challenges out there. In addition to having to hunt for every meal, servals are in danger of being poached and losing their habitat to humans.

SIAMESE

The **SIAMESE** breed originated in Siam, the country that is today called Thailand, where they are thought to have been the trusted and sacred guards of Buddhist temples. Their name in the Thai language is wichien maat, which translates to "moon diamond." With their bright, sparkling eyes, Siamese cats are certainly diamond-like, which is why they are such popular pets.

SIBERIAN

Gorgeous gentle giants, **SIBERIAN** cats are sometimes also called Siberian forest cats and Moscow longhairs. An adult male can weigh a whopping twenty pounds, making them a large cat breed. This powerful breed's coat has helped it survive the great outdoors during harsh Russian winters for at least a thousand years. They are known to be adventurous and easygoing. They have larger paws than similar breeds, such as the Maine coon and the Norwegian forest cat.

SINGAPURA

The **SINGAPURA** cat looks like a kitten no matter how old it gets. An adult male weighs eight pounds at most. Their coats are an unusual cat color: sepia brown. These cats look as though they sparkle in the sunlight thanks to multicolored stripes on each strand of hair. They have big, dramatic eyes that can be green, hazel, or yellow. Cats of this breed are extroverted and adventurous—you won't find them relaxing on couches for very long because they are always exploring their surroundings!

SNOWSHOE

This adorable blue-eyed breed looks like it's wearing snow-white mittens or shoes and originated from crossing Siamese and American shorthair cats. They are known to be mellow and sweet pets. **SNOWSHOE** cats are all white when they are born, but their brown markings slowly appear as they get older, though they always keep their white mittens.

SNOW LEOPARD

Nearly everything about the spectacular **SNOW LEOPARD'S** body has to do with helping it survive. Insulating fur keeps this big cat warm and dry in seriously snowy mountain environments. Its huge paws work like snowshoes, flattening out to keep it from sinking into the snow-covered ground. A long tail helps with balance, but also serves as a warm wrap on the coldest nights. Even its nose acts as a heater, warming the air before it gets to the snow leopard's lungs.

This leopard's heavy spotted coat has another important purpose: camouflage. These rugged cats are famously hard to spot. Snow leopards live high in the mountains in many central Asian countries, from Afghanistan to Nepal and far north into Russia. They can travel twenty-five miles in a single night looking for prey. Their favorite foods? Wild goats with huge horns (called ibex) and hairy goatlike animals (called Himalayan tahrs). Snow leopards also will eat marmots (a type of large squirrel), hares, rodents, and birds.

Only a few thousand snow leopards remain in the wild. Despite their remote residences, poachers hunt these magnificent cats to sell their fur. Some herders kill snow leopards that threaten their livestock (farm animals). Scientists, conservationists, and cat lovers around the world are working hard to protect this sensational species.

SOKOKE

The **SOKOKE** is among the rarest of rare domestic cat breeds. It has a ticked pattern all through the darker areas on its coat that likely helped its wild ancestors hide from predators. This sweet, short-haired breed is native to Kenya. It is named after the Arabuko Sokoke Forest, a protected area found on the coast of that country.

SOUTHERN ONCILLA

The dense jungles of South and Central America are home to predators of all shapes and sizes—from big cats and big birds to toothy predators that lurk in murky waters. Others, like the **SOUTHERN ONCILLA**, come in a cuter package. This small but mighty hunter weighs up to five pounds and preys on rats, mice, birds, and lizards. Also called southern tiger cats, these fierce felines spend a lot of time hiding from bigger predators so as not to get eaten themselves.

SPHYNX

You might think the **SPHYNX** cat originated in Egypt, given its name. But the first sphynx was born in Canada in 1966, a wrinkly kitten who was appropriately named Prune.

These unusual-looking cats have a thin, downy coat that feels like peach fuzz. They have doglike personalities: friendly, clownish, and loveable. Sphynx owners who take their cats outside sometimes dress them in UV-blocking cat shirts or apply pet-friendly sunscreen to prevent them from getting sunburns. During winter, these hairless kitties need a sweater on chilly days to keep them warm.

SUNDA CLOUDED LEOPARD

The **SUNDA CLOUDED LEOPARD**, which lives on the islands of western Indonesia in a region called Sundaland, is a separate species from the clouded leopard on the mainland of Malaysia. The Sunda clouded leopard has darker fur and smaller spots with a similar cloudy look. One thing that's weird: Neither of these species is actually that closely related to leopards at all.

SUNDA LEOPARD CAT

Indonesia is home to the **SUNDA LEOPARD CAT**, a genetically different species than both the much larger leopard and the mainland leopard cat. The Sunda leopard cat is a nocturnal animal who prefers to hunt on the ground but sleep in the trees. The Sunda leopard cat plays an important role in its island environment: helping to keep rodent populations in check. It also likes to eat lizards, birds, snakes, and frogs.

THAI LILAC

The first **THAI LILAC** kitten naturally occurred in a litter of Korat kittens when one special cat came out with a much lighter, purple-tinted coat, thanks to a genetic mutation. Thus, the breed came to be. The Thai lilac has points (darker markings on its face, ears, legs, paws, and tail) that have a purplish tinge. These people-oriented cats make for very friendly pets. They love going for car rides. Thai lilac cats form tight bonds with their humans and are very loyal. They have huge eyes that take up most of their faces.

TIGER

The **TIGER** is an awe-inspiring and remarkably beautiful animal. Each tiger's stripes are unique, like an individual work of art. Stripes twist, merge, and mingle, making shapes like diamonds, ovals, letters, and parallel or intersecting lines.

The world's biggest cat, the tiger is a massive predator built to kill large prey. It can weigh up to six hundred pounds—as much as a grizzly bear. One paw might be six inches across, and from nose to tail, its body can stretch up to nine feet long. Tigers have lived in the dense forests of Asia and Russia since prehistoric times. Survival adaptations such as powerful jaws, massive teeth, and large claws help them survive.

There are six endangered kinds of tigers, including the Bengal tiger, the Siberian tiger, the Sumatran tiger, the Indochinese tiger, the Malayan tiger, and the South China tiger. There are also three extinct tiger subspecies: Caspian tigers, Bali tigers, and Javan tigers.

Tigers exude power, but they are also the wildcat closest to extinction. A century ago, about a hundred thousand tigers lived free. Today, there are more tigers in captivity than in all the world's shrinking tiger habitats. Fewer than six thousand remain in the wild around the world. But there's good news: In some areas, tigers are making a comeback. Southern India now supports the largest wild tiger population in the world with about 3,600 of them. These tigers live alongside wild Asian elephants as well as an abundance of tasty prey animals such as sambar deer, spotted chital deer, and wild pigs.

TURKISH ANGORA

This diminutive, silky-soft breed is considered a national treasure in Turkey. They originated in the capital city of Ankara, which used to be called Angora. Some call these graceful kitties the ballerinas of the cat world. They are very outgoing and love to be the center of attention. Unlike many cats, **TURKISH ANGORA** kitties love to splash and play in water.

TURKISH VAN

The **TURKISH VAN** is known for its distinctive colored ears and tail. These cats have blue or gold eyes—and sometimes they even have one of each. The Turkish Van is also known for its ability to swim. It is from a rugged region centered around the largest lake in Turkey: Lake Van. This kitty loves heights and can often be found gazing down on its people from whatever shelf it was able to climb up onto.

TURKISH VANKEDISI

The silky white **TURKISH VANKEDISI** shares the same history as the Turkish Van. The name Vankedisi translates to "cat from Van" in Turkish. These smart kitties are playful and especially enjoy a game of fetch. While Turkish Van cats have white bodies with splotches of different colors on their ears and tails, Turkish Vankedisis are completely white. Many of these cats have heterochromia, meaning they have one blue eye and one green eye.

YORK CHOCOLATE

This delicious-sounding breed is reported to have been developed by accident on a goat farm in New York in the early 1980s. Fluffy fur in shades of black and brown are the most noticeable characteristic of these cats and gave them their name. Most of these cats are a solid dark chocolate color. When a **YORK CHOCOLATE** is a bit lighter in color, it is referred to as lavender. These cats are excellent mouse hunters, but they also have a soft side and love to be lap cats when they aren't on the prowl. In Italy, where these cats became popular, they were nicknamed ii gatto cioccolato—which means "the chocolate cat."

RESOURCES

Built on the long legacy of TIME, TIME for Kids has been a trusted news source in schools for 30 years, providing educators with valuable resources for the classroom. From articles about new scientific breakthroughs to profiles on inspiring kids who are helping their communities, TIME for Kids has content to inspire every reader.

Learn more about cats and other interesting animals with TIME for Kids!

Visit **www.timeforkids.com** and explore the **Animals** topic to discover your next favorite species.

EXPLORE A BIG CAT SANCTUARY NEAR YOU!

Black Beauty Ranch—Murchison, Texas

Black Pine Animal Sanctuary—Albion, Indiana

Carolina Tiger Rescue—Pittsboro, North Carolina

Cedarhill Animal Sanctuary—Caledonia, Mississippi

Exotic Feline Rescue Center—Center Point, Indiana

Forest Animal Rescue—Silver Springs, Florida

In-Sync Exotics Wildlife Rescue and Educational Center—Wylie, Texas

Popcorn Park Animal Refuge—Forked River, New Jersey

Safe Haven Wildlife Sanctuary—Imlay, Nevada

The Roar Foundation/Shambala Preserve—Acton, California

The Wild Animal Sanctuary—Keenesburg, Colorado

The Wildcat Sanctuary—Sandstone, Minnesota

Turpentine Creek Wildlife Refuge—Eureka Springs, Arkansas

WildCat Ridge Sanctuary—Scotts Mills, Oregon